John B. Hicks

An Inaugural Dissertation on the Compression of the Brain
from Concussion

Submitted to the public examination of the faculty of physic, under the

authority of the trustees of Columbia College in the state of New-York :

William Samuel Johnson, LL.D.

John B. Hicks

An Inaugural Dissertation on the Compression of the Brain from Concussion
*Submitted to the public examination of the faculty of physic, under the authority of
the trustees of Columbia College in the state of New-York : William Samuel Johnson,
LL.D.*

ISBN/EAN: 9783337373511

Printed in Europe, USA, Canada, Australia, Japan

Cover: Foto ©berggeist007 / pixelio.de

More available books at **www.hansebooks.com**

A N

INAUGURAL DISSERTATION

O N

COMPRESSION OF THE BRAIN FROM CONCUSSION.

SUBMITTED TO THE PUBLIC EXAMINATION

OF THE

FACULTY OF PHYSIC,

UNDER THE AUTHORITY OF THE

TRUSTEES OF COLUMBIA COLLEGE

IN THE

STATE OF NEW-YORK:

WILLIAM SAMUEL JOHNSON, LL.D. Prefident;

FOR THE DEGREE OF

DOCTOR OF PHYSIC;

ON THE THIRTIETH DAY OF APRIL, 1793.

By JOHN B. HICKS,

Citizen of the State of New-York.

Experience and Obfervation, the Parents of the Healing Art.

LEDRAN.

NEW-YORK:

Printed by T. and J. SWORDS, Printers to the Faculty of Phyfic of Columbia College, No. 27, William-Street.

—1793.—

Imprimatur.

Richard Bayley.

MEDICAL SOCIETY

STATE OF *NEW-YORK:*

GENTLEMEN,

PERMIT me to addrefs this INAUGURAL ESSAY, not as worthy of your fanction, but, as an evidence of efteem, refpect, and gratitude.

From your much obliged,

Humble Servant,

The AUTHOR.

ISAAC ROOSEVELT, Prefident,

THEOPHILACT BACHE, Vice-Prefident,

JOHN MURRAY, Treafurer,

JOHN KEESE, Secretary,

HENRY HAYDOCK,
WILLIAM EDGAR,
SAMUEL FRANKLIN,
THOMAS PEARSALL,
JOHN MURRAY, Junior,
GEORGE BOWNE,
LAWRENCE EMBREE,
JACOB WATSON,
SAMUEL JONES,
THOMAS BUCHANAN,
GERARD WALTON,
HUGH GAINE,
ALEXANDER ROBERTSON,
ROBERT BOWNE,
RICHARD MORRIS,
JOHN LAWRENCE,
ROBERT R. LIVINGSTON,
JAMES WATSON,
CORNELIUS RAY,
MATTHEW CLARKSON,
SAMUEL OSGOOD,
MOSES ROGERS,

GOVERNORS OF THE NEW-YORK HOSPITAL.

GENTLEMEN,

Your liberal exertions in fupport of the NEW-YORK HOSPITAL demand my approbation as a citizen, and gratitude as a child of the inftitution.

Your judicious felection of MEDICAL OFFICERS not only effectuates the public bounty to thofe victims of difeafe who are objects of the inftitution,

tution, but enables the patrons of the HEALING ART to eſtabliſh one of the firſt PRACTICAL MEDICAL SCHOOLS on the Continent of America.

You have combined in the ſame body, a ſource of balm to the afflicted, and obſervation for the advancement of the moſt noble of ſciences; and thus the remoteſt parts of the habitable globe will participate the fruits of your benevolence.

As men ſtudious to promote the arts and ſciences, permit me to ſuggeſt the flouriſhing ſtate of MEDICAL EDUCATION in this city. An inſtitution recently adopted by COLUMBIA COLLEGE, nearly allied to, and cheriſhed by the one under your direction, though in its infancy, ſupported by the Pillars of Literature, is now emerging from obſcurity, and will ſoon ſhine conſpicuous by its tendency to diffuſe USEFUL KNOWLEDGE.

How pleaſing the reflection, that, in this inſtitution, COLUMBIA boaſts of a youth of genius and erudition, unequalled in the hiſtory of ſcience.*

May PROVIDENCE continue to be your guide, and your future adminiſtration be ſuch as to merit a continuance of the confidence of thoſe whom you have the honour to repreſent.

I remain,

With much eſteem and reſpect,

Your obliged,

And very humble Servant,

The AUTHOR.

* Profeſſor MITCHILL.

INTRODUCTION.

INTRODUCTION.

THE *Senforium Commune has attracted the atten-*
tion of the ableft Philofophers. Its phyfiology has eluded
the refearches of ingenuity and induftry, is involved in
obfcurity, and hence the uncertainty of the pathology.
We have to lament the impoffibility of exploring this hid-
den, important, and inimitable machine of INFINITE
WISDOM.

Though Surgery has rapidly improved for a few years
paft, ftill, within its province exift a variety of difeafes;
the caufe, nature, and cure of which baffle the ingenui-
ty of Surgeons.

Although the obfervations which I now fubmit to
public fcrutiny conftitute my INAUGURAL ESSAY, *they*
were not compofed merely in compliance with the STA-
TUTES *of the inftitution under whofe fanction I publifh,*
but from a full conviction of their propriety; for though
they have been fubjects of invcftigation, fince Surgery has
been cultivated as an art, we are in poffeffion of no fatis-
factory information refpecting them.

In

In profecuting thefe fubjects, an interefting queftion offers for critical examination; to wit, Are the difeafes known in furgical authors, by concuffion and compreffion of the brain, effentially different ?

It is of infinite moment in practice to have difeafes defined with accuracy; for though modern induftry has unburthened Surgery of its ancient barbarity, yet, in many inftances, we are under the difagreeable neceffity of obtaining the aid of painful and dangerous operations; and hence the queftion ftated has not only agitated and perplexed our predeceffors, but the ableft of our cotemporaries are embarraffed, and have fplit on this ground.

Though I am bold to relinquifh the beaten tract, and controvert the theory and practice of eminent Surgeons, I prefume I fhall not be impeached with a thirft for novelty or a fpirit of controverfy. If I have advanced opinions repugnant to eftablifhed practice and great authority, it has been with all deference; from a conviction of their propriety, and a defire of alleviating the calamities attendant on mortality.

The ignorant, illiberal, and rafh may charge me with vain prefumption; but the wife, candid, and deliberate will applaud my independence.

This difparity of fentiment is not peculiar to the healing fcience; but we are not equally interefted and

folicitous

ſolicitous for the purity of each: On one the health and lives of ſociety are depending; the others are more the inquiries of curious ſpeculation than real utility. I am an advocate for lenity; but when I contemplate the nature and importance of the ſubjeʓ, I ſpurn at the degeneracy of authors.

The imperfeʓ ſtate of this part of ſcience has ariſen, in ſome degree, from its intricacy, ſervile and ſuperſtitious veneration for authority; but depravity of principle has had its influence; and thus the light of truth has been obſcured by the ſhadow of the imagination; and to eſtabliſh a favourite hypotheſis, faʓs have been coined, and wanton cruelties praʓiſed.

We are reſtrained by humanity, and prohibited by the laws of civil ſociety, from ſporting with the calamities and lives of the human creation; and hence the propriety of the obligation makes every conſideration ſubordinate to the welfare of the patient. *This, however humiliating, and painful to the reflecʓion, is too frequently abandoned, and intereſt and ambition predominate.*

INAUGURAL DISSERTATION, &c.

IT is almoſt unneceſſary to remark, that this diſeaſe is not peculiar to any ſeaſon or climate; every
part of the habitable globe, both ſexes, all ages,
and every variety of conſtitution are the objects of
its vengeance. A complete or imperfect, general
or partial ſuſpenſion of the intellectual operation,
ſenſe, and voluntary motion, from external violence,
with the continuance of reſpiration and circulation,
is a definition ſufficiently clear and diſtinct, and
plainly enough directing the indications of cure.
But in obedience to cuſtom, and further, becauſe
the variety marked has contributed to give riſe to
an opinion and method of cure which, in the ſub-
ſequent part of this eſſay, I ſhall take the liberty to
controvert; I proceed to mark, with preciſion, the
hiſtory in particular.

When

When the violence is inflicted the following is the order of symptoms.

The person being thrown to the ground, lies apparently lifeless to the most accurate examination. The body, face, and extremities are pale and cold; the eyes lose their sprightliness, become languid, and assume a deadly appearance; the action of the heart and arteries, and the function of respiration, are not perceptible, or extremely obscure.

In this situation he continues for a longer or shorter time, and too frequently ends his existence: But generally the involuntary organs recover their suspended operations.

The breathing is deep, sonorous, laborious, and slow; the pulse irregular, labouring, and oppressed; but, sometimes the lungs play with freedom and ease, the pulse is free, soft, regular, and full; and, in short, the state of the *vital functions* corresponds to that of *sleep*.

The face is flushed, and often livid; the eyes turgid, and the pupils dilated; the stomach is affected with nausea, sickness, and vomiting; intestines constipated; and these symptoms are frequently accompanied with an effusion of blood from the nose, ears, and eyes.

The

The muscles, subservient to the will, are variously convulsed; particularly the muscles of one side are violently agitated with alternate contractions and relaxations, while those of the opposite are in a state of perfect paralysis.

These convulsions are often extremely obscure, the muscles being affected simply with tremor, especially manifest in the distortion of the features of the face.

Thus situated for an unlimited time, suspended between hope and despair, we have no certainty of approaching death, nor evidence to warrant a conjecture of probable recovery.

At length the symptoms are mitigated, the breathing becomes more free and easy; the pulse regular, full, more quick, and frequent; sensibility in some degree returns; he is restless, awake to active stimuli, is sensible to light; the pupils contract, the eyes exhibit the appearance of intoxication, " and he talks incoherently."

Thus, gradem per gradem, he verges from the jaws of death, again to exercise the functions of life, and exhibit health, that inestimable blessing. Some degree of fatuity continues for a short time; and we have instances of its continuing for life, accompanied with paralysis.

Our

Our efforts to obtain this falutary termination are frequently baffled; the difeafe purfues a different courfe; and inftead of the fymptoms laft detailed, thofe of an oppofite complexion take place. The breathing is fmall, flow, and very obfcure; the patient is abfolutely infenfible; his pupils remain fixed and dilated when expofed to the ftrongeft light; his face and extremities are pale and cold, pulfe fmall, very flow, weak, and intermittent; the urine and fæces pafs involuntarily; and now all hopes of cure are precluded, and death is at hand. At length general convulfions fupervene and conclude this painful fcene.

REMOTE CAUSES.

AUTHORS have generally confidered the divifion of the remote caufes into occafional and predifpofing, as fuperfluous; becaufe, (fay they) the exciting never requires the aid of predifpofition, and the difeafe always arifes from their power alone.

That this is generally the cafe will not be denied. But, I will rifk an opinion, that inftances have occurred where the difeafe has been avoided from the abfence of predifpofition, and in every occurrence is aggravated thereby.

The

The neceſſity of this further appears, and is particularly enforced by its influence in determining the
prognoſis, and directing the indications of cure.

OCCASIONAL CAUSE.

COMMOTION or concuſſion of the brain.

PREDISPOSITION.

THIS I hold to conſiſt in a phlethoric ſtate of the
ſyſtem in general, and of the head in particular.
General phlethora may ariſe from original ſtamina,
a liberal uſe of animal food and ſpirituous liquors;
frequent intoxication, ſedentary life, imbecility in the
intellectual operation, interruption or ſuppreſſion of
the ſecretions, ſuppreſſion of accuſtomed evacuations;
and, we are told that frequent bleeding will have the
ſame effect.

Phlethora of the head, in particular, may depend
on original conformation.

A large head, ſhort neck, contracted thorax, and
a conſtitutionally leſs capacity of the lungs in proportion to the other parts of the body.

Corpulent

Corpulent habit, offification of the valves of the heart, and debility of the fame.

Caufes referred to the lungs, varioufly interrupt-ing, impeding, and retarding the minor circulation.

Effufions of blood, ferum and extrication of air in the cellular fubftance of the lungs.

Hydrops pericardii, hydrothorax, fpafms, rigi-dity, and tremors; tympanites, aneurifms, afcites, amputation; youth, old age; and, in fhort, what-ever will increafe the determination to, or prevent the return of blood from the head, will produce a phlethora of the fame.

PROXIMATE CAUSE.

THE queftion which I am now to difcufs, is as interefting to the Patient, intricate and perplex-ing to the Practitioner, as any within the jurifdiction of Surgery.

Surgeons have agreed on the caufe, nature, and cure of compreffion of the brain from external vio-lence; imputing the difeafe to compreffion, depend-ing on extravation, congeftion, depreffion of bone, and the introduction of foreign fubftances: But
the

the fancy of modern inquifitors has invented a new fpecies of difeafe, entitled Concuffion of the Brain; though of a nature diametrically oppofite, in the character fo nearly allied to compreffion, that the diftinction would elude the penetration of a *Newton.*

With all deference to fuch authorities, I believe the fame to have no exiftence in nature, and to be a phantom of the imagination only.

This diftinction was firft fuggefted by the French,* who are unbounded in ambition, enthufiafts in novelty, happy at invention, and fond of the marvellous.

Like contagion it has diffufed its noxious influence, affimilated the Surgeons of the adjacent and remoteft countries; and to this caufe too many owe their diffolution.

Authors of a more recent date, unfatisfied with the obfervations of their predeceffors, have refined on refinement; the former having denied the exiftence of compreffion in concuffion, to which the latter have added debility as the proximate caufe, and attributed to concuffion a direct fedative operation, analogous to contagion, cold, fear, and other known fedatives.† What a ftretch of the imagination!

C I have

* Sharp's Critical Inquiry.
† Bell's Syftem of Surgery.

I have attentively weighed, and revolved in my mind, in the moft favourable point of view, the arguments urged in fupport of concuffion independent of compreffion, and I am forry to fay, I have not been able to difcover a rational foundation; and no doubt exifts with me as to the fallacy of their obfervations.

Upon the whole, I prefume I fhall clearly maintain, that the variety of the phænomena marked in the hiftory of compreffion, and other circumftances; on which the doctrine of concuffion has been founded, may be referred to a difference in the predifpofition, &c. which gives no effential difference, implies an analogous operation in the occafional caufe, marks a variety only, and fhews the deception of experience.

Though I have rejected the idea of excluding compreffion in what authors denominate concuffion, I admit that debility is induced, is to be fought for to explain many phænomena, and is of infinite moment in the method of cure.

This debility I hold to be indirect, the refult of exceffive ftimulus, and not the immediate effect of direct fedative powers.

From

From this confideration only we can explain the inftantaneous fufpenfion of life, and immediate reftoration of the fame, when the brain is fubjected to violent concuffion.

That debility, and not compreffion, is the caufe of the phænomena in this inftance, is obvious from the following reflections.

What reafon can be given why the functions of life, after a complete fufpenfion, inftantly return to exercife with their ufual energy?

Abforption is a tedious, laborious procefs, and no one acquainted with the laws of the *animal œconomy* can conceive extravafation and abforption in the inftance mentioned ; and it is equally improbable that congeftion exifted.

All caufes, fo far as they have a mechanical operation, are, to the human body, direct ftimulants. By comparing the phænomena of concuffion with the morbid effects of electricity, joy, anger and pain, the moft perfect analogy muft ftrike the unprejudiced and enlightened mind.

As the phænomena of concuffion have the ftricteft refemblance to thofe of electricity, &c. it is unphilofophical

lofophical to fay, that the fame effects refult from oppofite caufes.*

Having refuted the idea of direct, and eftablifhed, upon invulnerable ground, the exiftence of indirect debility, it remains to fhew the exiftence of com-preffion in concuffion.

This I conceive a difficult tafk, more fuited to age, experience and obfervation, than calculated as an exercife for a youth yet an infant in his profeffion.

Neceffity needs no apology, and I affume the ar-duous undertaking with that diffidence becoming my period of life.

I have already maintained the exiftence of debility refulting from concuffion; it may feem paradoxical now to urge the neceffity of compreffion.

I wifh to be clearly comprehended. The opinion I have already advanced has a relation to circum-ftances, feldom the objects of furgical obfervation.

No one is ignorant, that inftances daily occur, where life is fufpended by falling from a height on the feet, head, or other parts; and on the moft accu-rate examination, we have no evidence of folution of continuity, contufion, extravafation, or congeftion.

We

* " Identity of effect implies identity of caufe." .

We have a more familiar example of this in the impious and shameful practice of boxing. We have all witnessed, and many have experienced what pugilists consider as an invariable rule, that a blow under the ear, between the mastoid process of the temporal bone and angle of the jaw, throws the victim to the ground, where he lies apparently inanimate; but in a few moments he is revived, renews the contest, and engages his antagonist with redoubled energy.

This state of concussion is not what Surgeons are contending about. When the *intellectual faculties*, *sense* and *voluntary motion* are no more, and the *vital functions* continue to display themselves; when these exist to a considerable degree, and are of some duration, they constitute what is to many simple concussion, and to others complete compression.

I hold it as a principle in *physiology*, that a free and uninterrupted state of the *nervous power*, (whatever that may be,) is necessary to the existence and exercise of the *intellectual faculties*, *sense* and *voluntary motion*.

It is equally true, that the *animal functions* may be suspended, and the *vital* continue in action. It appears, therefore, that a cause capable of effecting the first, may be inadequate to the second; and that in the disease now under consideration, though an

injury

injury of some kind has arrested the *animal*, the *vital organs* are supported by that state of the *sensorium*, or *nervous energy*, which, though insufficient for the former, is equal to the latter. I am aware of the objection, that the phænomena which I have just laid down as depending on compression, are frequently the effects of a state of the *nervous system*, in which we have, (say they) no cause to suspect compression.

The first of which is *sleep*.

That this state is not produced by extravasation is obvious; but to me it is equally probable, that, though compression is not the first cause of sleep, it completes and preserves that state, and without which it would never take place,

I am persuaded of the truth of this from a variety of considerations,

The mechanism of the brain, which is calculated to retard and accumulate the blood in this organ, while no other use, equally probable, can be assigned, (except giving a certain degree of energy to the *nervous system*)* is sufficient evidence, in a case intricate as the present, to warrant the assertion.†

" Beside

---

* On one, the most simple principle, nature accomplishes two of the most important ends in the animal œconomy, EXCITEMENT and COLLAPSE.

† This explanation of the SENSORIUM does away what has hither-to been considered a DESIDERATUM in PHYSIOLOGY; to wit, a reason

" Befide the caufes now mentioned occafioning
apoplexy by compreffion, I alledge there are other
caufes producing the fame difeafe, by directly deftroying the mobility of the nervous energy: Such
caufes feem to be mephitic air arifing from ferment-
ing

why a greater quantity of blood is diftributed to this organ than is neceffary
for its nourifhment. On this has been built the favourite hypothefis of
the NERVOUS FLUID. If we reafon from analogy, (fay they;) the brain
is a GLAND, deftined for the fecretion of a peculiar fluid; it being a fact,
that, whenever a greater quantity of blood is fent to any organ than is required for its fupport, it is intended for fecerning from the circulating
mafs a certain fomething SUI GENERIS. To this rule there are objections;
the SPLEEN, PANCREAS, and CAPULÆ RENALES; in none of which
have we any evidence of their glandular ftructure, no EXCRETORY DUCT
having been difcovered; nay, I may with propriety fay never will.

The explanation given is confiftent with the more remote caufes of fleep.
Exercife of the various functions of the body, by exhaufting the energy
of the fyftem, favours congeftion in the veffels of the brain; for, as the
veffels of this part forward their blood by their own energy and the vifa-
tergo; fo, in cafes of general debility, parts thus fituated, are moft rea-
dily furcharged; and hence the HEAD, THORACIC, and ABDOMINAL
VISCERA are moft frequently affected with congeftion. In this way Nature compels the moft obftinate to yield, and enforces recreation. We
have a further illuftration of this from the fomniferous effects of cold.
We all know, that when expofed to a degree of extreme cold, we become
drowfy; and, fenfible of the impending evil, in vain we refift, fleep ap-
proaches, and, if the caufe continues, we expire without a fenfe of pain,
anxiety, or regret. I conceive that cold produces thefe effects, by diminifhing the energy of the fyftem; the heart, unable to exert its ufual force,
propels the blood with difficulty in the extreme parts, and, therefore, it
accumulates internally, but more particularly in the veffels of the brain,
for the reafons before given.

A difficulty here arifes—if congeftion is the caufe of fleep, what folution will be given, why a perfon can be inftantly awoke? Compreffion
from extravafation, and that from congeftion in the veffels, are different
with refpect to duration. In the firft, compreffion can only be relieved by
abforption, while, in the latter, a reftoration of action alone is all that is
requifite; therefore, on the caufe of excitement being given, he is awoke;

ing liquors, and from many other fources; the fumes
arifing from burning charcoal; the fumes of mercury, of lead, and fome other metallic fubftances;
opium, alcohal, and many other narcotic poifons:
To which I would add the power of cold, of concuffion, of electricity, and of certain paffions of the
mind.

" None of thefe poifons, or noxious powers, feem
to kill by acting on the organs of refpiration, or
fanguiferous fyftem; and I believe their immediate
and direct operation to be on the nervous power,
deftroying its mobility; becaufe thefe poifons fhew
their power in deftroying the irritability of mufcles,
and of the nerves connected with them, when both
thefe are entirely feparated from the reft of the body.

" With refpect, however, to the circumftances
which may appear on diffection of perfons dead of
apoplexy,

but even this is not immediately, for frequently he continues in a ftate of
femi-watching for fome time; and when he is revived, on withdrawing the
caufe, he finks in the fame ftate; and if he does not, it is only after
fome time that the watching ftate is completed.

I conceive, however, that the caufes principally to be noticed in explain-
ing the phænomenon of fleep, is a peculiar organization and habit. We
have a beautiful illuftration of this in the organ of voice: here the moft
aftonifhing phænomena in the animal machine depend on peculiarity of
organization, and a facility of action in the mufcles fubfervient, acquired
by repetition: fo the fame caufes referred to the SENSORIUM, may produce
congeftion and inanition not to be expected a priori; and hence the caufe
of children fleeping fo much more than adults, and as they advance to
puberty of its gradually diminifhing.

apoplexy, there may be some fallacy in judging from
those of the cause of the disease. Whatever takes off,
or diminishes the mobility of the nervous power, may
very much retard the motion of the blood in the vessels of the brain, and perhaps to a degree of increasing
exhalation, or even of occasioning rupture and effusion; so that, in such cases, the marks of compression
may appear upon dissection, though the disease had
truly depended on causes destroying the mobility of
the nervous power. This seems to be illustrated
and confirmed from what occurs in many cases of
epilepsy: in some of these, after a repetition of fits,
recovered from in the usual manner, a fatuity is induced, which commonly depends upon a watery in-
undation of the brain; and in other cases of epilepsy,
when fits have been often repeated without any per-
manent consequences, there happens at length a fatal
paroxism; and, upon dissection, it appears that an
effusion of blood had happened. This, I think, is
to be considered as a cause of death, not a cause of
the disease; for, in such cases, the disease had dimi-
nished the action of the vessels of the brain, and
thereby given occasion to a stagnation, which produced the appearances mentioned. And, I apprehend the same reasoning will apply to the retro-
cedent gout, which, by destroying the energy of the
brain, may occasion such a stagnation as will produce

D   rupture,

rupture, effufion, and death; and in fuch cafes the appearances might lead us to think that the apoplexy had depended entirely on compreffion."

In thefe quotations we have the authority of Doctor Cullen, together with extreme plaufibility, to fhew, that concuffion, though it produces debility in the nervous fyftem, is attended with compreffion.

This the Doctor will not admit as a caufe of the difeafe, but as an effect of death, for what reafon I am at a lofs to fay.

I might with equal propriety contend, that extravafation or depreffion of bone from external violence, were not the caufes of an apoplexy in fuch inftances, becaufe they were occafioned by external violence; and, therefore, the force applied is the caufe, and extravafation or depreffion of bone, the effect of death: For what is the difference, whether the effufion or extravafation is the effect of rupture by violence, or rupture and effufion from a want of energy in the veffels?

In the former the apoplexy evidently depends upon extravafation and depreffed bone; and it is equally true, that, in the latter, though debility of the nervous energy occafioned the rupture and effufion, debility independent of that would not produce the

apoplectic

apopleÅic ftate; for, if, it would, a reftoration of
energy would be the cure: this, is contrary to faÅ,
and therefore debility is the remote, and compreffion
the proximate caufe, and not the effeÅ of death.

The faÅ which I am now to lay down as invariably
the cafe in every inftance, and for the truth of which I
appeal to every PraÅitioner experienced in Surgery,
that when concuffion of the brain is of duration
fufficient to fall under the obfervation of Surgeons
and Phyficians, it will be found, that the *animal* and
vital funÅions were extinguifhed at the inftant of
concuffion, and that this is momentary: For though
the *animal funÅions* continue in a ftate of fufpenfion,
the *vital* renew their aÅion; and in this ftate the
perfon remains for an unlimited time.

If concuffion, independent of compreffion, fup-
ports the difeafe under the circumftances I have mentioned, I afk a reafon why the *animal* as well as the
vital funÅions were not reftored.

A reply to this queftion I conceive will be difficult, and a rational folution only can be given on the
principle of compreffion.

That debility alone does not exhibit the phænomena, I conclude from hence, that if it did, there would

be

be no paufe at the *vital*, and the *animal functions* would be reftored in the fame progreffion: This, however, is otherwife; for the *animal functions* do not proceed to a reftoration of action in the fame progreffion, but remain in a ftate of abolition for a confiderable time, and at length frequently end in death.

This is a ftubborn objection to the doctrine of concuffion independent of compreffion, and we muft purfue fome other courfe to arrive at the truth. I would attempt it in this way: Violence of *excitement* from concuffion terminates in a ftate of *collapfe* analogous to *fyncope* from joy,

I confefs, that, in this ftate, the neceffity of compreffion does not appear; I prefume at the fame time, that extravafation accompanies the excitement, or, that congeftion in the veffels takes place at the time of *collapfe*.

If *fyncope* from joy does not terminate in a ftate of irrecoverable *collapfe*, it in a few moments difappears, and excitement returns; fo, in cafe of *collapfe* from concuffion, the fame takes place, except with this difference, that in *fyncope* excitement is reftored to the whole fyftem; whereas, in concuffion, to the *vital functions* only.

The

The following is the reafon.

I have ftated it as a principle in *phyfiology*, that a caufe, capable of fufpending the *animal*, may be inadequate to produce the fame effect on the *vital organs*: Now, I conceive, that in the inftance of concuffion, life is fufpended from *collapfe* only, and the excitement is reftored to the *vital functions*; becaufe extravafation or congeftion in the veffels producing compreffion, exifts at the fame time; and though inadequate to any manifeft injury on the *vital*, is competent to a lofs of the *animal functions*.

Bell, in his Syftem of Surgery, fays, " So far as my obfervation goes, the moft material difference which occurs between the fymptoms produced by thefe two caufes, concuffion and compreffion, is met with in the pulfe and in the breathing.

" In a compreffed ftate of the brain, the breathing is commonly deep and oppreffed, fimilar to what moft frequently takes place in apoplexy; whereas, in patients labouring under commotion or concuffion, the breathing is in general free and eafy, and the patient lies as if he was in a found and natural fleep. The pulfe is commonly foft and equal, and not irregular and flow, as it ufually is found to be when the brain is compreffed.

" In

" In cafe of compreffed brain too, although little or no relief may be obtained from blood-letting, yet no harm is obferved to accrue from it; for in fuch circumftances it may be prefcribed in moderate quantities, without reducing either the frequency or ftrength of the pulfe; whereas, in real concuffion of the brain, the pulfe, as we have already remarked, will frequently fink, and become more feeble on the difcharge of only eight or ten ounces of blood."

Thefe obfervations, on the circumftances of the two difeafes, may induce an inattentive obferver to adopt the diftinction; but, I apprehend I fhall be able to account for the variety quoted, and admit the exiftence of compreffion.

There is a ftate of the arterial fyftem which *pathologifts* have named *phlethora ad vires*; and we all know the exiftence of *inanition* produces an oppofite ftate; the compreffion being given, thefe two ftates of the fanguiferous fyftem will give the variety: For inftance, if a *phlethora ad vires* fhould exift in one, and *inanition* in another, the firft will prefent the fymptoms of compreffion; the fecond will exhibit the phænomena of concuffion.—In fupport of this an explanation may be required.

The

The pulfe in *phlethora ad vires* is irregular, flow and oppreffed, and correfponds to that of compref-fion. This does not arife from a want of energy in the *nervous fyftem*, but from the equilibrium be-tween the *arteries* and their contents being deftroyed, from the predominancy of the latter.

The fame confideration accounts for the deep and oppreffed breathing; for, as this *organ* is a mixture of voluntary and involuntary, and as the *animal functions* are fufpended, the *mufcles*, fubfervient to the organ of refpiration, refufe their affiftance; and hence the ftate of refpiration in compreffion. Further, the blood is accumulated in the right *au-ricle* and *ventricle* of the *heart*, from the flownefs of refpiration, the diminifhed energy of the *heart* and *phlethora*; and, therefore, a deep infpiration is ne-ceffary to the tranfmiffion of the accumulated blood.

The breathing is free and eafy, the pulfe foft and equal in concuffion, and anfwers to that of fleep.

I have no doubt but Practitioners have been de-ceived by thofe circumftances of the breathing and pulfe, becaufe I fhall now prove them to depend up-on compreffion.

The cavity of the *crarium* being given, and always the fame, at different times it will be more or lefs

completely

completely filled, according to the state of the *vessels* of the *brain*. If two persons of unequal states of *phlethora* should have accumulated in the scull the same quantity of matter, one may receive compression equal to a complete state of *apoplexy*, and the other shall receive no injury, or if he does it will be in a less degree; for the accumulation being given, the compression will be in proportion to the *phlethoric state* of the *brain*.

A more inconsiderable accumulation and state of *inanition* of the *vessels*, will explain why a soft, equal pulse, and a free, easy respiration takes place in this instance and not in the other. As respiration is in some degree under the influence of the *will*, it will be impaired in proportion to the loss of the same; and as the degree of respiration required will be in proportion to the quantity of blood to be circulated by the *lungs*, we have a reason, without the necessity of a comment, why the state of the *vital organs* corresponds to a state of *sleep*, having previously shewn this phænomenon to depend on compression.

Dissection, so frequently the resource of Surgeons, is often fallacious.

We have it roundly asserted in books of surgery, that after the most attentive examination of many
who

who died of concuffion, they were not able to difcover the leaft veftige of compreffion.

It would be illiberal to charge the authors of
fuch diffection with wilful mifreprefentations of the
refult of their inquiries; but, I may with the
ftricteft propriety obferve, that it did exift in every
fuch inftance, though it eluded the eyes of thofe
blinded by prejudice.

I fhall anticipate the reply to this obfervation, that
the fame refult has attended the purfuits of the
warmeft advocates for compreffion: I admit the
fact, but deny the force of this as an objection.

Compreffion from external violence has hitherto
been confidered as produced by extravafation, depreffed bone, and the introduction of foreign fubftances: was this a fact, I fhould be in fome meafure
defeated by the energy of their diffections: but,
compreffion from concuffion is frequently the effect
of congeftion in the veffels of the brain.

Again, though extravafation fhould exift in thofe
inftances in which we are told diffection has dif-
proved, I fay, I can eafily conceive, that the moft
accurate and liberal diffectors may have been de-
ceived; for the extravafated matter neceffary to
produce this effect under particular circumftances,

E   is

is too inconfiderable to ftrike the fenfes, unlefs the
fame be fubjected to the moft fufpicious examination.

The circumftance I have juft mentioned is not
the only caufe of error in fuch cafes; for the feat
of compreffion has generally been confidered as
exifting between the *duramater* and *crarium*, the
former and *pia mater*, or the latter and *brain:* I
conceive, however, that extravafation may with
equal facility take place in the fubftance of the
brain, or fome other internal part, as thofe for-
merly mentioned.*

Further, extravafation of blood is not a neceffary
confequence of rupture of the veffels of the *brain* from
concuffion; for ferum only may be, and frequently
is effufed, and proves a caufe of compreffion.

Now, from the facts which I have juft mentioned; to wit, a falfe idea of the feat of compreffion,
a partial knowledge of the caufes, and from the
inconfiderable quantity of extravafated matter neceffary to produce this effect, it is probable, that
diffection has done little more than nourifh an abfurd preconceived opinion.

As congeftion in the veffels of the *brain* is the
caufe in many inftances of compreffion from concuffion,

* Pott's Surgery.

fion, the reafon is obvious, why diffectors have been fo often baffled in difcovering a caufe of compreffion.

If they had no fufpicion of the exiftence of fuch a caufe, is it probable that if it did exift, it would fail under their obfervation? I anfwer no; becaufe the nature of congeftion may be, and generally is fuch, as to render its exiftence ambiguous to thofe who contend for its prefence,

The celebrated Bromfield and others found, from long extenfive experience and much obfervation, that the *afthenic* plan was too rigidly obferved; and therefore, without a knowledge of the caufe of the inefficacy of their practice, *empirically* ventured to relinquifh the fame, and fubmit to *fudorifies* in fuch cafes.*

I have no doubt that the indifcriminate practice of *venefection* in concuffion is often injurious; and I am equally certain, that *fudorifies*, *cathartics*, and every other part of the *afthenic* plan, are no lefs ambiguous under particular circumftances.

I am led to believe, both from *theory* and *practice*, that the fatality of this difeafe is owing to the prejudice in favour of fyftem, whereby, practifing from general indications, many and infuperable errors are committed.

If

* Bromfield's Surgery,

If the mortality of this difeafe has abated fince Surgeons and Phyficians have fufpected the propriety of *bleeding* in every inftance, and all ftages of the difeafe, we have reafon to conclude, that, if for the *afthenic* they had fubmitted the *fthenic*, the proportion of deaths in fuch inftances would have much diminifhed.

I am at a lofs to fay, why the author of the prefent Syftem of Surgery has advanced the experience and obfervations of Bromfield, in fupport of his ridiculous hypothefis on the proximate caufe of concuffion.* I admit that it is evidence of the exiftence of debility; but this I fhall fhew is not a caufe, but a neceffary confequence of that ftate of the *brain*.

The exercife of the *intellectual faculties, fenfe, voluntary motion* and *refpiration*, are direct ftimulants to the fyftem, and the only caufe and fupport of its activity.

It is now a notorious fact, that *refpiration* is a procefs analogous to combuftion; and that the office of the *minor circulation* is to impregnate the blood with what, in the new nomenclature of the French chemifts, is known by *oxigene*, and to prove the fource of *animal beat*.

It

* Bell's Syftem of Surgery.

It is equally well eſtabliſhed, that the exiſtence and exerciſe of this *function* is eſſential to animal life: it appears further, that the preſence of *oxigene* is neceſſary to the action of the *heart* and *arteries*, it being the natural and only ſtimulus capable of ſupporting the ſame.

Now, whoever will reflect calmly on what has been ſubmitted on the *animal* and *vital functions*, may eaſily conceive from the ſubduction of ſtimulus, that in every caſe of compreſſion of any duration, debility muſt be the reſult.

In this way I preſume the diſeaſe proves fatal.

In all inſtances of inanition, exiſting at the time of extravaſation or congeſtion, in conſequence of concuſſion, I ſay, bleeding is an ambiguous remedy, becauſe the indication is to take off compreſſion by the removal of extravaſated or infarcted matter: this can only be effected by abſorption and a reſtoration of energy to the infarcted veſſels.

A queſtion here ariſes, Is veneſection adequate to, or compatible with the indication?

This is a ſubject equal to an Inaugural Eſſay, and improper to be diſcuſſed at preſent; I ſhall only remark, that in my opinion it is.

The

The fuccefs of Bromfield's practice appears to be the refult not of judicious prefcriptions, but a leffer degree of improper practice.

We adduce, from their own experience, the moft invincible evidence of the non-exiftence of debility, as the proximate caufe.

All the champions for this favourite hypothefis fpeak favourably of the ufe of *cathartics*, as affording fingular relief.

It is almoft unneceffary to remark, that the energy of the *brain* depends upon a certain ftate of tenfion in the veffels of the fame, as is obvious from depletion producing *fyncope*: examples of which we have in the removal of a ligature from the arm in venefection, *paracentifis thoracis et abdominis, and parturition.*

The indication of cure, (agreeable to Bell and others) is to *invigorate* the *fyftem.*

Who can reflect on fuch fingular inconfiftency without difguft!

What are the effects of *cathartics?* To debilitate.

How far they are calculated to fulfil the indication needs no comment.

I prefume we are now in poffeffion of an incontrovertible fact, to fubvert the ftately edifice of brilliant and fertile imaginations.

Does

Does not the invariable and fuccefsful practice of *purging* in *phrenitis*, *fynocha*, and all other *fthenic* difeafes, and the general averfion of Phyficians to their exhibition in the *afthenic*, fpeak a plain language; to wit, that their falutary effects in concuffion can never be reconciled to debility as a caufe, and, therefore, is one of many in proof of the non-exiftence of the fame?

From the preceding reflections on the proximate caufe, it muft be obvious, that the theory of the fame is the following.

When the brain is fubjected to violent commotion, the fyftem is greatly excited; this terminates in a ftate of collapfe analogous to fyncope from joy, and generally is of fhort duration; for the perfon in a few moments recovers his ufual excitement: But fometimes the *vital functions* only recover their action, and the *animal* continue in a ftate of fufpenfion. In this ftate of the difeafe, I prefume indirect debility to have no agency in producing the phænomena, and that the difeafe is fupported by compreffion: Again, that direct debility arifes from the fubduction of ftimulus, and in this way the difeafe proves fatal.

METHOD

METHOD of CURE.

WE may amuse our fancy, and indulge our ingenuity in visionary speculations on the *theory* of *diseases*; but, the interposition of *art* in administering relief, is of too serious a nature for such exercise.

As I have dared to relinquish the beaten path, and propose a change in practice, it may be considered necessary that I should, in conformity to custom, shew the fallacy and inefficacy of the *systems* of our predecessors and cotemporaries on this subject.

The limited nature of this essay renders such an undertaking impracticable: I presume, however, that the arguments I shall adduce in support of my own, will sufficiently evince the impropriety of those which have preceded.

The observations which I have previously submitted on the subjects of concussion and compression, leave no doubt as to the nature of the proximate cause; and therefore, the indication resulting therefrom is obvious; but to effect the same is ambiguous and difficult.

The indication is to take off compression.

As

As compreſſion in this inſtance is produced and
ſupported by congeſtion in the veſſels of the *brain*,
or extravaſation, we can only remove the immediate
cauſe by *abſorption*, or a reſtoration of *energy* to
the veſſels.

It would be of advantage if we were in poſſeſſion
of circumſtances to enable us to diſtinguiſh ſuch
caſes as are occaſioned by *extravaſation*, from thoſe
depending on *congeſtion* in the veſſels ; but as this is
impoſſible, and as three fourths of the whole depend
on *extravaſation*, we muſt chooſe the leaſt of two
evils—abſorption therefore is what we have in view.

As I have confined my inquiries to that variety
of compreſſion which ariſes from concuſſion or commotion, no indication can preſent with reſpect to the
operation of the *trepan*.*

F   The

---

* We are directed by Bell, in his Syſtem of Surgery, in all caſes of
COMPRESSED BRAIN, though we are in poſſeſſion of no circumſtances
to direct to the ſeat of the injury, to TREPAN every acceſſible part of the
CRANIUM.

If ever a thirſt for novelty led a man into a labyrinth of folly, we have
an inſtance of it here. Would not a Phyſician be impeached with inſa-
nity, or charged with ſporting with the life of his patient, who, in caſes of
APOPLEXY, ſhould direct the uſe of the TREPAN ? and is not the proximate cauſe of APOPLEXY the ſame as that of compreſſion from concuſſion ? If ſo, have we any one fact, which in compreſſion can require or
juſtify an operation, more than in APOPLEXY ? If the operation of the
TREPAN was unattended with pain, and of ſuch a nature that we need
apprehend no danger from its uſe, I ſay, under ſuch circumſtances, if the
probability of relieving the cauſe of compreſſion was not more than one of

The remedy in which I repofe the greateft confidence, to effect the indication, is *venefection*.

Could I reconcile the idea, that depleting the *fanguiferous fyftem* will increafe and invigorate the action of the *abforbent*, I fhould with eafe furmount every difficulty as to the propriety of the remedy: But, as I am not in poffeffion of facts for its fupport, and as I have no defire to impofe, by fubtilty of reafoning, on the minds of men, at the expence of my patients, I muft relinquifh the idea, as wild, vifionary and abfurd; although adopting it would greatly aid to enforce the fyftem of practice which I wifh to eftablifh.

It may be afked, what are the circumftances of this difeafe, which indicate the ufe of the *lancet?*

To this I reply, that *venefection* is practifed on a principle different from what it is in inftances of *phlegmafia*; in the latter we have in view the production of debility, while in the former, my object is to deplete until the fymptoms of compreffion vanifh without inducing debility.

Whether

a thoufand, the operation would be warranted : But, as the operation is tedious and laborious to the Surgeon, and painful and dangerous to the patient, it is injudicious, wicked and cruel, and will be avoided by all wife, prudent and honeft Practitioners.

Whether the patient is of *plethoric* habit or not, blood should be drawn at certain intervals, until symptoms of relief appear, or death ensues; for the extravasation can only be removed by *absorption*; and as this is a tedious operation, the patient may die from the subduction of stimulus.* Further, as the *vital* is in some measure under the influence of the *animal functions*, and these are in a state of suspension, *venesection*, independent of relieving the patient from the danger of subduction of stimulus, by the same operations, is calculated to promote absorption.

In performing this operation, the following circumstances should be strictly observed:—The blood should flow from a small orifice, so that the stream may not exceed the diameter of a hair: this should be interrupted on the discharge of an ounce, by closing the orifice, and repeated every ten minutes until one of the effects I have mentioned appear.

The situation of the patient should be horizontal, because this will counteract the tendency to a *deliquium animi.*

During the intervals of bleeding, the energy of the system should be supported by the *diffusible* stimuli;

* Browne's Elements of Medicine.

muli; the exhibition of which muſt be directed by
the judgment of the Practitioner.*

I am ſenſible of the inconveniences which will
reſult from any increaſe of debility, and I have ac-
cordingly directed the evacuation to be made in a
manner, which, I conceive, will effect the removal
of phlethora, without the acceſſion of debility in
proportion to the quantity loſt: For two pounds diſ-
charged in the way I have preſcribed, will not be
equal to three ounces from a large orifice, without
interruption, and favoured by an erect poſture.

I truly lament the neceſſity of an ambiguous remedy, but as we are in poſſeſſion of none ſo proba-
bly judicious, the difficulties attending its uſe muſt
be diſpenſed with: The objections, though ſpecious,
will in a great degree yield to a ſcrupulous examination.

I have obſerved, that the *vital* are in ſome degree
influenced by the *animal functions:* Now, as the latter are in a ſtate of ſuſpenſion, the abſorbents will
be diminiſhed in proportion to their dependence on
the ſame.

Again,

* Theſe will counteract the enervating tendency of veneſection, and
thus preſerve, under the evacuation, the vigour of the ſyſtem.

Again, as the cavity of the *cranium* is always the same, and the bulk of the *brain* in proportion to the state of its veffels, I fay, it is obvious from thefe confiderations, that, as the extravafation neceffary to effect compreffion is very inconfiderable,* *vene-fection,* by contracting the *fenforium*, will take off compreffion; confequently, fo far as the energy of the *abforbents* was impaired by this caufe, it will be renewed; and the recovered fufpended *functions* will afford feveral fources of ftimulus; therefore, the evacuation, fo far from enervating, invigorates the fyftem.

The ftate of the *inteftines* fhould be early attended to; for by difcharging the contents of thefe, the *phlethora* of the head will be relieved by deriving to the inferior parts: This intention will be beft effect-ed by ftimulating *cathartics*, the beft of which is the *mercurius dulcis*; for it not only is attended with this laft effect, but, by a *ftimulus* given to the abforbents of the *inteftines*, communicated by confent to the
remainder

---

* I was prefent at an operation of the TREPAN, performed by that ju-dicious Phyfician Doctor William Moore, of this city : in this inftance, the patient laboured under a complete compreffion of the BRAIN, and the extravafated matter did not exceed five grains; on the removal of which he inftantly recovered his fenfes.

remainder of the fame fyftem of veffels, promotes abforption.*

When the animal functions are in fome meafure reftored, all further evacuation fhould be prohibited, and the cure fubmitted to mercury: But as a recovery of *phlethora* previous to the removal of extravafated matter would occafion a return of the difeafe, this fhould be avoided by due attention in preferving a favourable balance between the *ingefta* and *excreta.*

The fecondary fymptoms fo frequently occurring and generally fatal, are attributed to a variety of caufes. Mr. Bell has fuggefted, that matter extravafated or effufed between the *cranium* and *pericranium*, by ftagnating becomes acrid, ftimulates and increafes the action of the veffels, and thus produces inflammation, which, by the communication of veffels, fpreads to the brain, and in this way proves fatal: I conceive that the fame extravafation, effufion, ftagnation and inflammation, may equally, and frequently does take place in the cavity of the *cranium*, and is attended with all the inconveniencies mentioned of the former. From this confideration, I
recommend

* On this principle I explain the effects of CATHARTICS in the cure of DROPSIES, and not by the evacuation giving occafion to abforption.

recommend the continuance of mercury for a confiderable time, even when all fymptoms of the difeafe
have entirely difappeared; and if this be ftrictly en-
joined, that train of fymptoms fo much feared,
often occurring, and in nineteen cafes of twenty ter-
minating in death, will be avoided.

F I N I S.